Discovering Whales and Dolphins

Written by Janet Craig

Illustrated by Pamela Johnson

Troll Associates

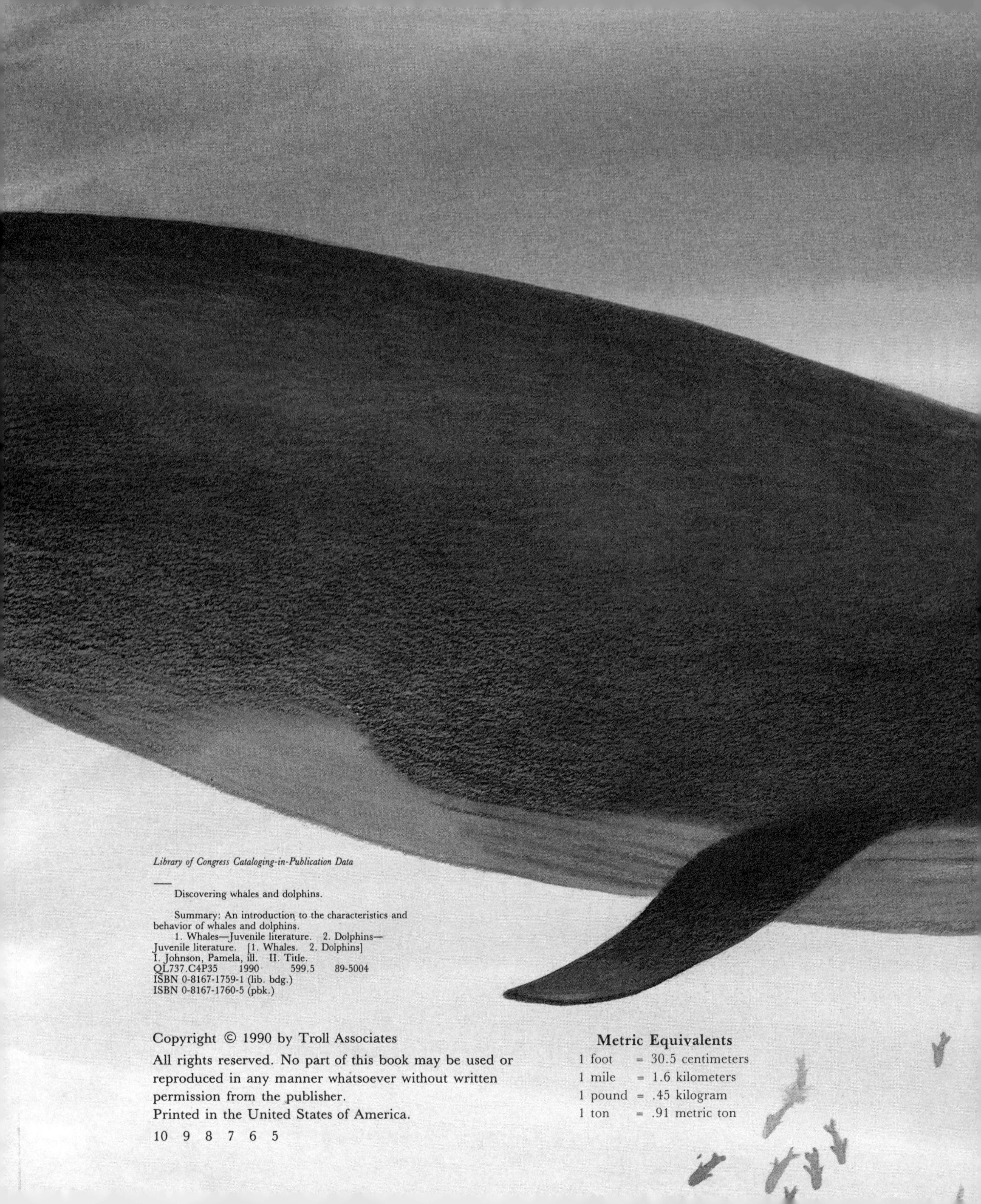

Library of Congress Cataloging-in-Publication Data

Discovering whales and dolphins.

Summary: An introduction to the characteristics and behavior of whales and dolphins.
1. Whales—Juvenile literature. 2. Dolphins—Juvenile literature. [1. Whales. 2. Dolphins]
I. Johnson, Pamela, ill. II. Title.
QL737.C4P35 1990 599.5 89-5004
ISBN 0-8167-1759-1 (lib. bdg.)
ISBN 0-8167-1760-5 (pbk.)

Printed in the United States of America.

10 9 8 7 6 5

Metric Equivalents

1 foot	=	30.5 centimeters
1 mile	=	1.6 kilometers
1 pound	=	.45 kilogram
1 ton	=	.91 metric ton

Imagine you are deep beneath the sea. Through the grayish-green water, fish dart about. Suddenly, a shadowy shape appears. It is huge! As it swims closer, you see what it is—a blue whale.

Gracefully, this bluish-gray giant dives in search of food. Then up to the top of the water it comes. A large cloud of steamy droplets fills the air as the whale breathes out. Taking in another big breath, the streamlined whale dives again.

The blue whale is the largest living creature in the world. It can weigh more than one hundred tons and measure more than one hundred feet long. That makes it as big as Ultrasaurus, one of the biggest dinosaurs that roamed the earth.

Blue whales, along with many other kinds of whales, live in all the oceans of the world. There are about seventy-five kinds of whales in all. They make up a beautiful and interesting group of animals.

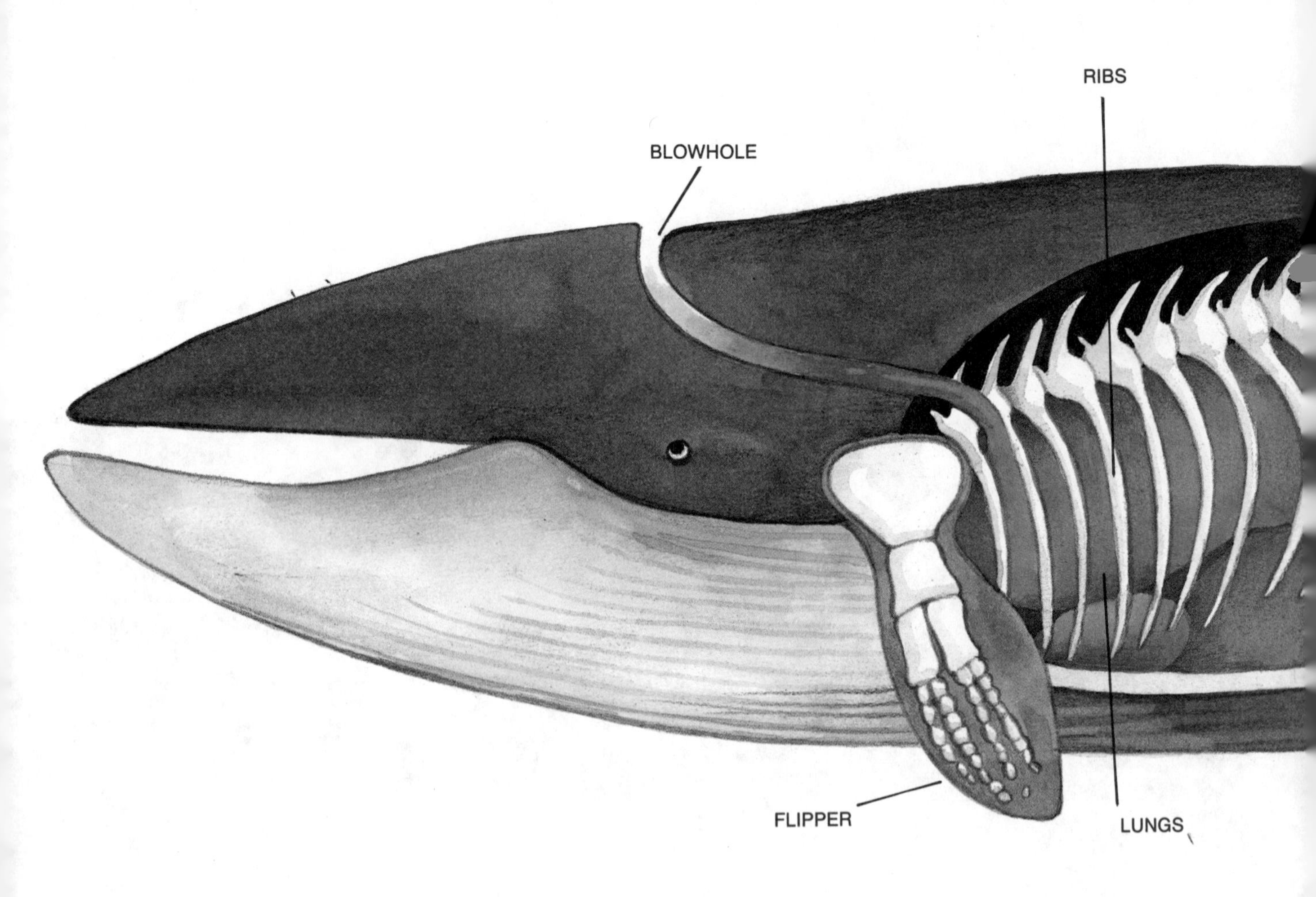

Whales spend their lives in water, but did you know that they are not fish? The whale is a mammal, just as you are.

Like other mammals, whales are warm-blooded. That means the temperature of their bodies stays the same, even when the temperature outside their bodies is colder or warmer.

Unlike fish, which have gills, whales have lungs to breathe air. That is why they must come to the top of the water to breathe. And although their skin is smooth to help them glide through the sea, some whales do have a few hairs on their heads.

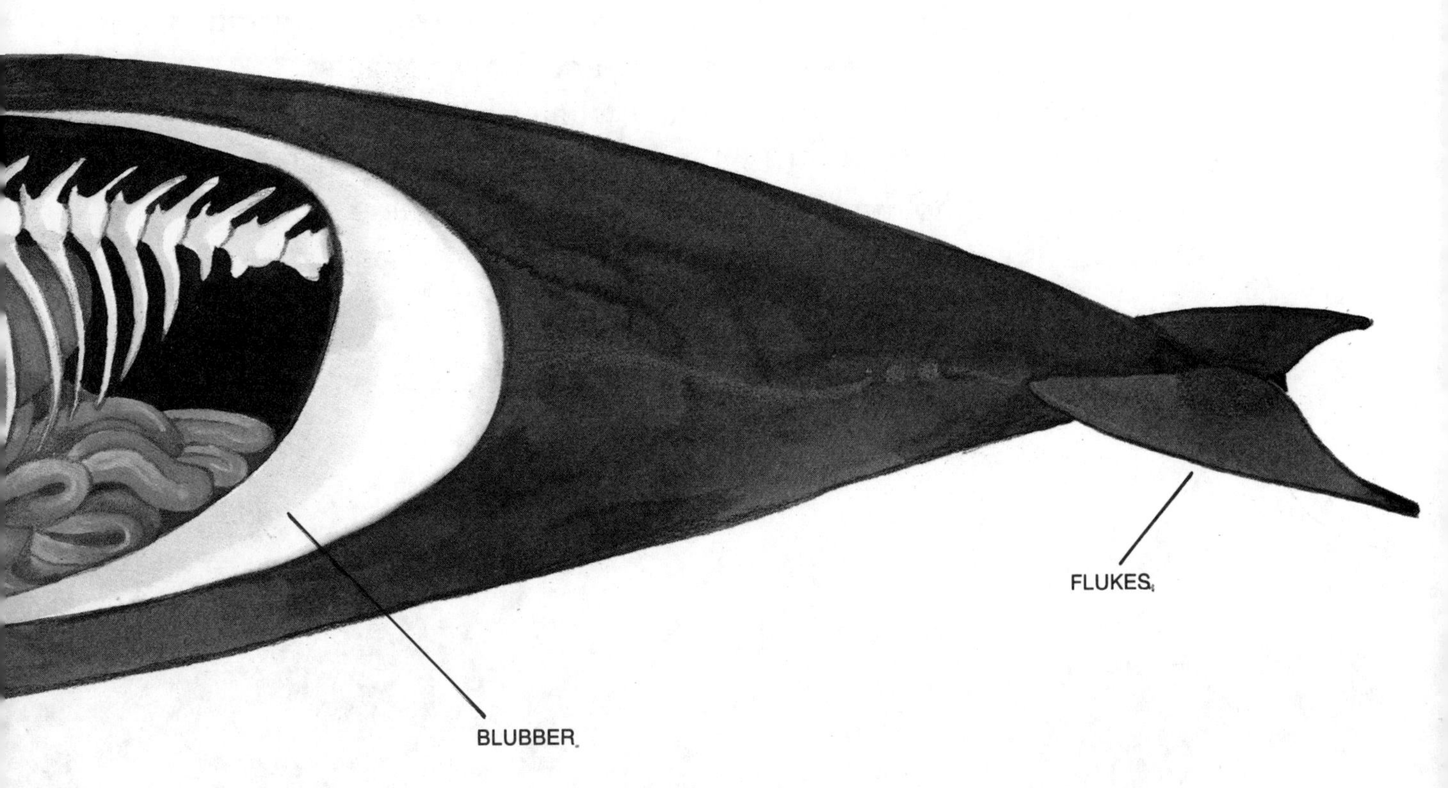

Scientists think whales first came from animals that were like deer or cattle. They had four legs, and they lived on land. Why did these ancient animals take to the water? No one knows for sure. But over millions of years, the whale's body changed to help it live in the sea. Its front legs became flippers to help it swim. Its hind legs disappeared. Instead, it grew powerful tail fins, called flukes, which move up and down in the water. This kind of movement helps the whale rise to the surface, then dive down again. The whale's body also grew a thick layer of fat. It is called blubber. Blubber keeps the whale warm in cold ocean water.

All whales are called *cetaceans*. This name comes from a word that means "large sea animal." There are two main groups of cetaceans. They are *toothed* whales and *baleen* whales.

Just as their name says, toothed whales have teeth. There are about sixty-five kinds of toothed whales. They use their teeth for catching and holding food, not for chewing. These whales like to eat fish and other sea animals.

The sperm whale is the largest toothed whale. Its big squarish head makes up one-third of its body. And inside that huge head is a twenty-pound brain! The sperm whale's head also has a valuable oil inside it, called *spermaceti*—and that is how this whale got its name.

The sperm whale swims along, seeking warm water. It dives very deeply—more than seven thousand feet. That is where it finds its favorite food, the giant squid.

A sperm whale can stay beneath the sea for more than an hour, waiting for a meal. Then up it swims. Its spout, a cloud of steam and droplets, fills the air like a giant fountain. The spout comes out of a special kind of nostril that whales have on their heads. This nostril is called a blowhole. Toothed whales have one blowhole, while baleen whales have a two-part nostril. To breathe, the whale uses strong muscles to open the blowhole. When the whale dives, the blowhole tightly snaps closed.

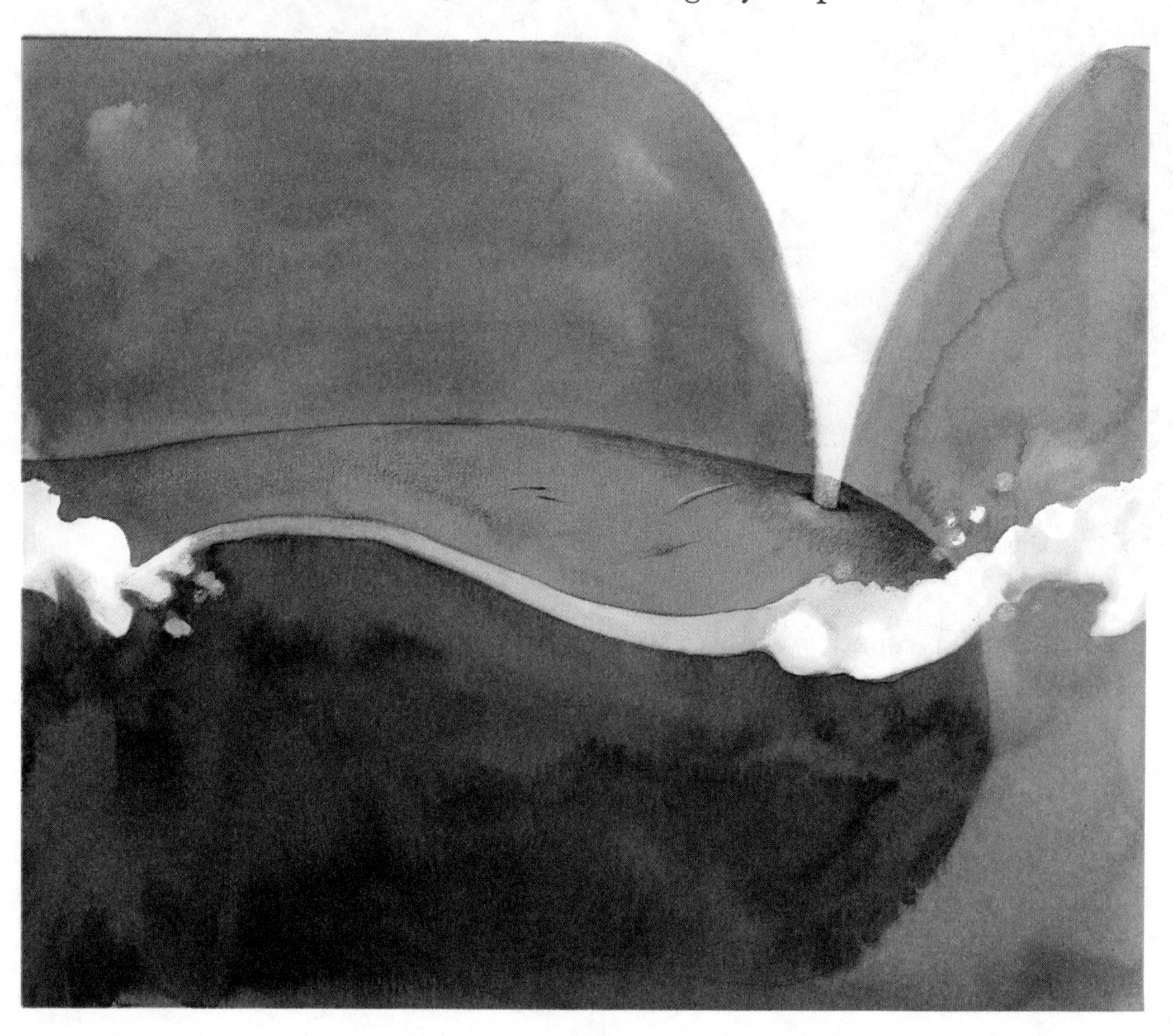

On a calm sea, a group of mother and baby sperm whales swim together. Suddenly, danger is near. A herd of killer whales is heading straight for the sperm whales.

Killer whales are another kind of toothed whale. Fearless killer whales will attack seals, dolphins, or even a giant blue whale! The killers' tall black fins stand straight out of the water as they fiercely swim toward their prey.

Swimming through the water at great speed, the killer whales try to attack a baby. Quickly, the mother whales go into action. They make a circle around the baby to protect it. This time the killer whales must move on. The baby is saved.

At other times, the killer whale spy-hops. It raises its head straight out of the water. With a sideways look, the killer watches a nearby ship.

Although they are feared by other sea animals, killer whales do not attack people. In fact, when they are captured, these whales seem very friendly. They like to learn tricks, and some will even let their trainers ride upon them.

Like most other toothed whales, killer whales are social animals. That means they live in groups and help one another during hunting and other activities.

The other group of whales, the baleen whales, are not social animals. Most baleen whales do not stay in large groups. But some, like the blue whale, swim in small family groups.

Baleen whales are toothless. They are named for the hundreds of long baleen plates that hang from the roofs of their mouths. Baleen, which is also called whalebone, is made of the same material as your fingernails. The edges of the baleen are fringed.

Baleen does a very important job for the whale. It strains food out of the water for the whale to eat.

What do baleen whales eat? Strangely enough, the biggest of all creatures, the blue whale, eats some of the smallest of creatures. These creatures are called plankton. They are the tiny plants and animals that float in sea water.

As the blue whale swims, its mouth opens wide. In goes a huge amount of plankton and water. Then the whale presses its great tongue against its baleen. The water is pushed out through the baleen's fringe, leaving a mouth full of plankton for the whale to eat. It takes a lot of plankton to satisfy this giant's appetite. A blue whale eats about four tons of food a day.

Each year, many baleen whales make long trips through the oceans of the world. These trips are called migrations.

The gray whale travels a very great distance. Its round-trip journey is about ten thousand miles. During summer, the gray whale feasts on the plankton found in Arctic waters. But when fall comes, these seas begin to freeze. It is time for the gray whale to start its journey south.

It swims and swims, until it reaches the warm waters of Mexico's Pacific Ocean. That is where the babies of gray whales are born. Then, when spring comes, the grays begin their journey back north.

Humpback whales also migrate. These baleen whales are easy to recognize. They have very long flippers, and there are bumps on their black heads.

Whale watchers love to see friendly humpbacks. With a powerful leap, a frisky humpback jumps out of the water. This is called breaching. Over and over, the giant jumps. What a beautiful sight it is!

Humpback whales are also known for their eerie but beautiful "songs." These special sounds have been studied by scientists. Only male humpbacks sing. Their songs can last up to thirty minutes and can be heard under water hundreds of miles away.

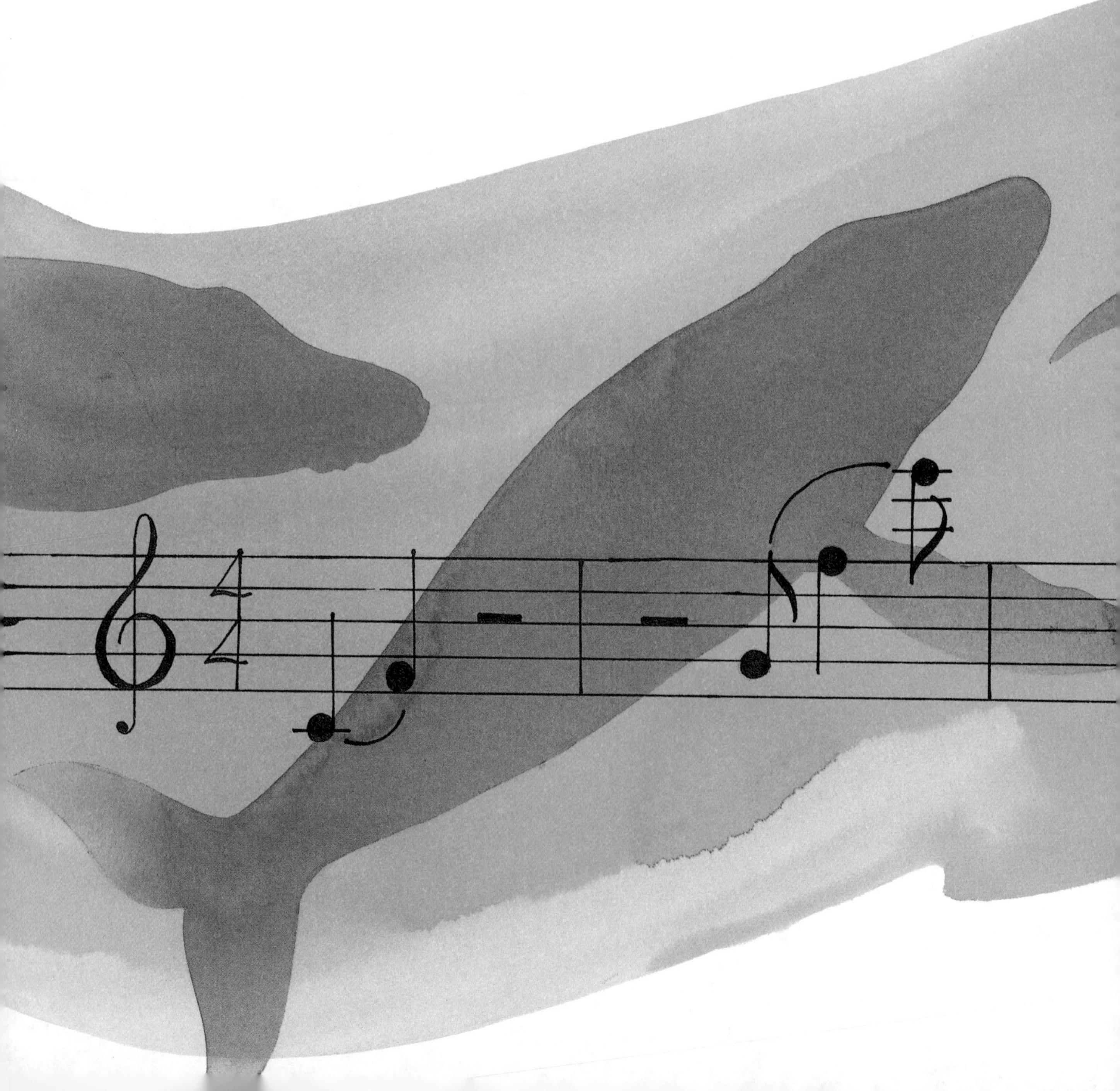

Why do humpback whales sing? Some people think the whales' songs may have to do with courtship, but scientists are not really sure.

Most whales are hard to study. Their large size makes it impossible for them to live in zoos or aquariums, where they could be watched closely.

However, there is a small toothed whale that we know a lot about. Perhaps you've seen it at the zoo, leaping high out of the water and performing amazing tricks. It is the bottle-nosed dolphin—a friendly looking sea mammal.

Some scientists believe dolphins are among the most intelligent of animals. In fact, bottle-nosed dolphins can be very creative. They will sometimes make up their own games, using floating rubber rings.

Dolphins are fun to watch. They seem to enjoy learning to jump through hoops, to grab objects with their mouths, and to catch and throw a ball.

Dolphins have excellent eyesight. But even in dark waters, dolphins and many other toothed whales have a special way to

"see" in front of them. It is called echolocation. Echolocation helps the dolphin to find food and to know if danger is near.

Here is how it works: As it swims underwater, a dolphin makes whistling or clicking sounds. These sounds bounce off objects in the water, making echoes. The dolphin listens for these echoes. From them, a dolphin can tell how close an object is and its size.

Whale watchers have also been able to see baby dolphins being born. The baby is born tailfirst. It is about three feet long and weighs twenty-five pounds. The baby, called a calf, keeps one flipper on its mother. Gently, the mother dolphin guides the calf to the top of the water for its first breath of air. Like other mammals, the baby drinks milk from its mother's body.

Although we understand some habits of dolphins and other whales, one activity is still a puzzle. It is called beaching.

At times, a whale or group of whales will swim to shore. They beach, or strand themselves upon the sand to die. Even when people guide the whales back to deeper water, they turn around and swim back to shore. No one knows the reason why whales do this. Perhaps they have a disease. Or perhaps a group of whales follows a sick whale to shore to try to help it.

Most whales are so large that they have few enemies in the sea. Perhaps their worst enemies have been humans. Over the years, people have hunted whales for their valuable oil, blubber, and baleen. So many whales were killed that certain kinds became scarce.

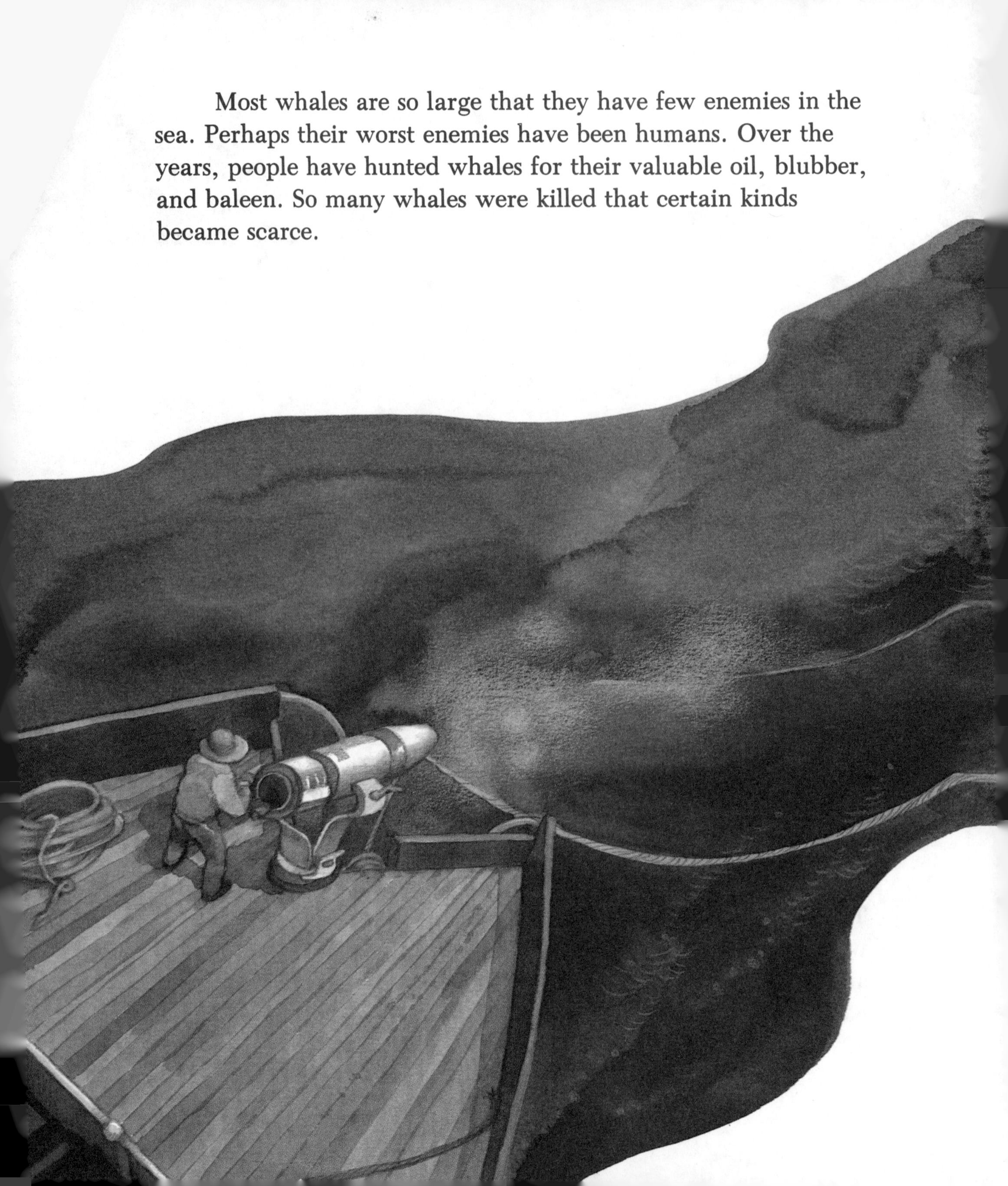

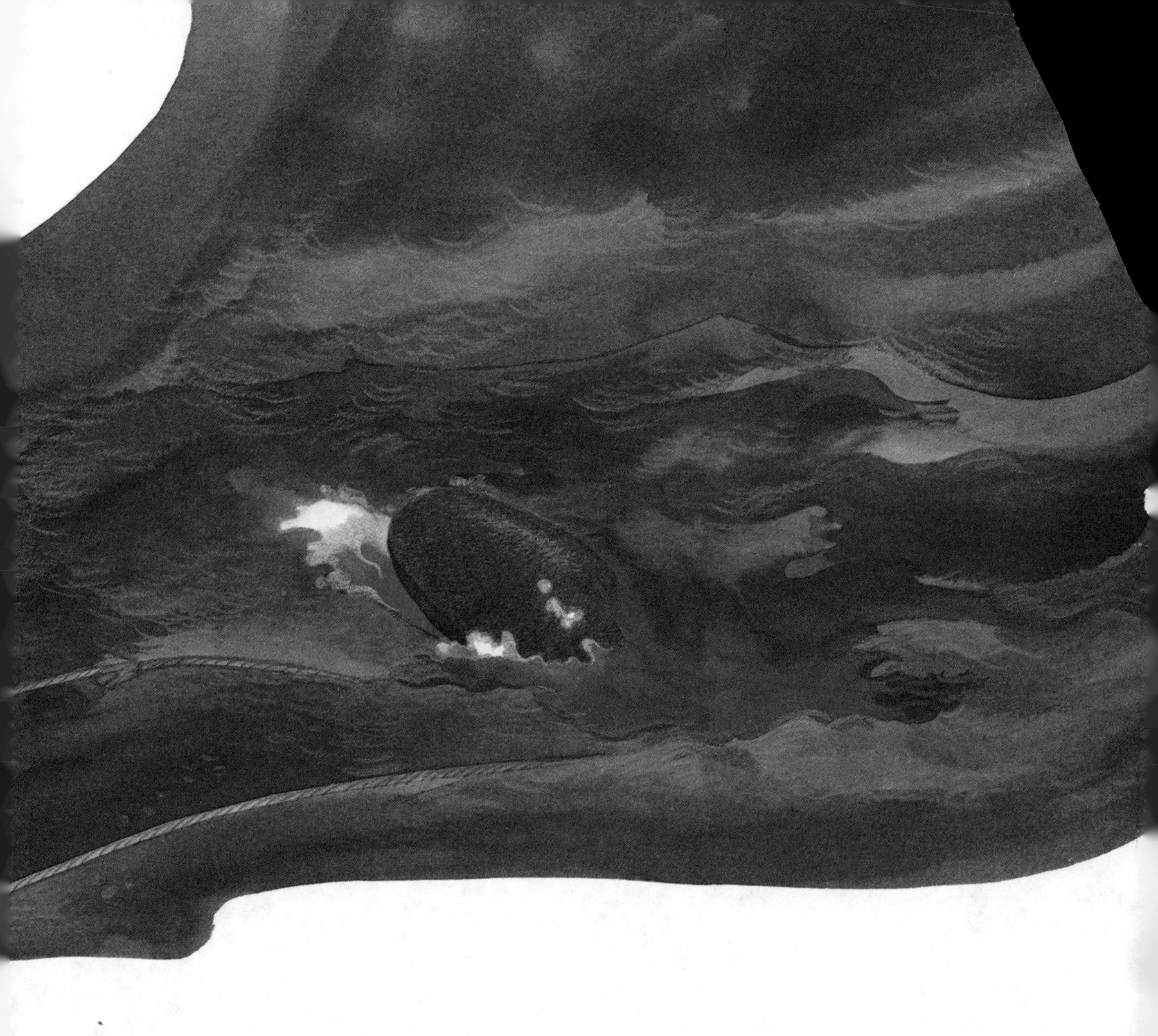

In 1946, many countries began to work together to protect the whales. They formed a group called the International Whaling Commission. So far, many whales have been saved. But the commission hopes to do more to protect these wonderful creatures.

Whales and dolphins make up an intelligent and very special group of animals. It is up to all of us to help preserve these beautiful creatures of the sea.